AF117420

THE POETRY OF COBALT

The Poetry of Cobalt

Walter the Educator™

SKB

Silent King Books a WhichHead Imprint

Copyright © 2023 by Walter the Educator™

All rights reserved. No part of this book may be reproduced in any manner whatsoever without written permission except in the case of brief quotations embodied in critical articles and reviews.

First Printing, 2023

Disclaimer
This book is a literary work; poems are not about specific persons, locations, situations, and/or circumstances unless mentioned in a historical context. This book is for entertainment and informational purposes only. The author and publisher offer this information without warranties expressed or implied. No matter the grounds, neither the author nor the publisher will be accountable for any losses, injuries, or other damages caused by the reader's use of this book. The use of this book acknowledges an understanding and acceptance of this disclaimer.

"Earning a degree in chemistry changed my life!"
— Walter the Educator

dedicated to all the chemistry lovers, like myself, across the world

CONTENTS

Dedication v

Why I Created This Book? 1

One - Element Of Grace 2

Two - Pigments To Alloys 4

Three - Mystical Allure 6

Four - Protector Of Life 8

Five - Cobalt, The Jewel 10

Six - Truest Power 12

Seven - Smartphones To Electric Cars . . . 14

Eight - Element Divine 16

Nine - Technology's Might 18

Ten - Day After Day 20

Eleven - The Enigma 22

Twelve - Marvel So True 24

Thirteen - A Wonder	. . . .	26
Fourteen - Cobalt, Oh Cobalt	. . . .	28
Fifteen - Source Of Inspiration	. . . .	30
Sixteen - Alloys And Superalloys	. . . . . .	32
Seventeen - Souls Take Wing	. . . .	34
Eighteen - Empowering Minds	. . . .	36
Nineteen - Force For Good	. . . .	38
Twenty - Progress Is Bred	. . . .	40
Twenty-One - Magnets To Batteries	. . . . .	42
Twenty-Two - Metal So Rare	. . . .	44
Twenty-Three - World Of Wonders	. . . . . .	46
Twenty-Four - Unlock The Potential	. . . . .	48
Twenty-Five - Twenty-seven	. . . .	50
Twenty-Six - Stars In The Night	. . . .	52
Twenty-Seven - Versatility Inspires	. . . . . .	54
Twenty-Eight - Cobalt's Essence	. . . . . .	56
Twenty-Nine - Year After Year	. . . .	58
Thirty - Force In Medicine	. . . .	60
Thirty-One - Brighter Future	. . . .	62
Thirty-Two - Every Single Hour	. . . .	64

Thirty-Three - Cobalt Ignites 66

Thirty-Four - Again And Again 68

Thirty-Five - Great Magnitude 70

Thirty-Six - Building Block 72

About The Author 74

WHY I CREATED THIS BOOK?

Creating a poetry book about the chemical element Cobalt was an intriguing and unique endeavor. Cobalt, with its distinctive properties and significance in various fields, offers rich material for poetic exploration. By delving into the characteristics, history, and applications of Cobalt, I can craft poems that celebrate its beauty, reflect on its role in human endeavors, or even use it as a metaphor for deeper emotions or experiences. This poetry book can educate, inspire, and captivate readers with its fusion of science and art, providing a fresh perspective on both Cobalt and the poetic form itself.

ONE

ELEMENT OF GRACE

In the depths of the earth's embrace,
A silent warrior takes its place.
A shimmering hue, a spectral gleam,
Cobalt, the element of dreams.

Born from the stars, forged in cosmic fire,
A testament to nature's desire.
A metal rare, with secrets untold,
It weaves its story, bold and bold.

In shades of blue, it dances free,
A symphony of light for all to see.
Cobalt, the artist's gentle hand,
Paints azure skies across the land.

In alloys strong, it lends its might,
To machines that pierce the darkest night.

Cobalt, the guardian of strength,
Defies the forces, at any length.

Yet hidden beneath its stoic guise,
A heart that beats, an alchemist's prize.
Cobalt, the healer, kind and pure,
Curing ailments with a gentle cure.

From batteries that power our days,
To pigments that grace an artist's gaze,
Cobalt, the element divine,
A treasure found within the mine.

So let us honor this noble guest,
With gratitude and deep respect.
For Cobalt, the element of grace,
Forever holds a special place.

TWO

PIGMENTS TO ALLOYS

In the realms of the periodic table, a tale unfolds,
Of a silent warrior, born from the stars, Cobalt, behold.
With shimmering hue and spectral gleam,
A mesmerizing element, unlike any dream.

    An artist it becomes, painting azure skies,
With its pigments, vibrant and wise.
Cobalt's brushstrokes adorn the world,
A masterpiece of beauty, gracefully unfurled.

    A guardian it stands, lending might to machines,
Powering the engines, fulfilling their dreams.
In batteries it resides, a source of endless power,
Igniting the flame, hour after hour.

    Yet Cobalt's touch goes beyond machines alone,
Its healing properties, a gift it has sown.

With gentle touch, it mends and cures,
A remedy for ailments, love it ensures.
    From ancient times to the present stage,
Cobalt has played a pivotal role on this earthly page.
From pigments to alloys, it weaves its spell,
A versatile element, in which wonders dwell.
    Oh, Cobalt, element of grace and might,
In the tapestry of nature, you shine so bright.
With every stroke and every touch,
You hold a special place, an element of much.
    So let us raise a toast, to Cobalt, our friend,
A silent warrior on whom we depend.
In its presence, we find solace and grace,
An element so precious, a gift to embrace.

THREE

MYSTICAL ALLURE

In the realm of elements, a silent warrior born,
From the cosmic forge, where stars are torn.
Cobalt, the guardian of machines and might,
With spectral gleam and shimmering hue, a stunning sight.

An artist it becomes, with a stroke of its brush,
Painting azure skies, with pigments that blush.
Vibrant and wise, its colors dance and play,
A masterpiece of nature, in every shade of blue and gray.

A healer it reveals, with a touch so gentle and pure,
Mending wounds and ailments, a remedy that's sure.
In the depths of its essence, secrets lie untold,
A potent elixir, a remedy to behold.

Cobalt, an element of grace and might,
A guardian, an artist, a healer in the night.

Let us raise a toast to its presence and worth,
For it blesses our lives, with its power and mirth.
 Oh Cobalt, we cherish your mystical allure,
Your touch, your hues, forever endure.
As we marvel at your wonders, so profound,
May your magic and brilliance always astound.

FOUR

PROTECTOR OF LIFE

Cobalt, oh Cobalt, your beauty astounds,
A metal of magic, with a mystical sound.
Born from the stars, a silent warrior you be,
Guardian of life, a healer, an artist, you see.

Your hue, a deep blue, like the ocean so vast,
A touch so gentle and pure, it's hard to surpass.
A protector of life, you keep us all safe,
A warrior of peace, with a heart full of grace.

Your brilliance shines through, like a star in the night,
A beacon of hope, with a guiding light.
An artist of nature, you paint a picture so fine,
A canvas of wonder, a sight so divine.

Cobalt, oh Cobalt, you're a treasure to behold,
A metal of wonder, with a story untold.

Your magic and brilliance, forever will stay,
A silent warrior, who brightens our way.

FIVE

COBALT, THE JEWEL

In the realm of art, a cobalt hue,
A pigment rare, vibrant and true.
Brush strokes dance, a vivid display,
Cobalt's allure lights up the way.

From canvas to pottery, its touch is found,
A whisper of magic, swirling around.
In deepest blues and shimmering greens,
Cobalt's essence, a masterpiece it weaves.

But Cobalt's power goes beyond the art,
Its healing touch, a balm to the heart.
A guardian of health, it's known to be,
A remedy for ailments, a remedy for me.

In the depths of the earth, Cobalt resides,
A power untamed, where secrets hide.
Its versatility, a marvel untold,
A force to reckon, a sight to behold.

Cobalt, oh Cobalt, your wonders unfold,
A symbol of strength, a story untold.
From paints to alloys, you shape our world,
A treasure of nature, in beauty unfurled.

So let us celebrate, this element divine,
Cobalt, the jewel, forever shall shine.
In art and healing, its legacy remains,
A testament to Cobalt's enduring reign.

SIX

TRUEST POWER

In the realms of the periodic table, behold Cobalt's might,
A guardian of strength, a healer in the darkest night.
Its essence, like a warrior, battles the tempest's rage,
An element of power, that never shall assuage.
    Cobalt, the silent sentinel, with steadfast grace it stands,
A myriad of secrets, hidden in its cosmic strands.
Its presence whispers tales of ancient alchemical lore,
Mysteries untold, waiting to be explored.
    A hue of azure, like the depths of the boundless sea,
Cobalt's brilliance shines, a beacon for all to see.
In artistry, it dances, a painter's precious tool,
Strokes of vibrant hues, creating wonders to enthral.
    Yet Cobalt's truest power lies in its healing touch,

A balm for weary souls, a salve that means so much.
It mends the broken spirit, brings solace to the heart,
A remedy for anguish, a cure for every part.

    Oh, Cobalt, noble element, guardian of life's symphony,
Your presence brings us solace, your essence sets us free.
In your enduring grace, we find strength to carry on,
Forever we shall cherish you, till the ages are long gone.

SEVEN

SMARTPHONES TO ELECTRIC CARS

In the realm of elements, Cobalt reigns,
A guardian of secrets, it sustains.
With hues of blue, a mystical shade,
Cobalt's prowess, never to fade.

A healer it is, in every way,
Mending ailments, night and day.
From heart to bone, it mends with care,
A touch of Cobalt, a cure so rare.

In art's embrace, Cobalt finds its place,
A painter's stroke, a sculptor's grace.
From vibrant ceramics to stained glass,
Cobalt weaves beauty, as time shall pass.

In every canvas, a story it tells,
A touch of Cobalt, the artist compels.

From Renaissance masters to modern dreams,
Cobalt's presence, like a silent stream.
  In alloys strong, Cobalt finds its might,
Forging a future, shining so bright.
From jet engines soaring through the sky,
To tools and turbines, Cobalt's supply.
  In batteries small, Cobalt sparks a spark,
Powering devices, even in the dark.
From smartphones to electric cars,
Cobalt's energy, reaching the stars.
  Oh Cobalt, element of wonder and grace,
With healing touch and artistic embrace.
You shape our world, in every way,
Cobalt, forever, we shall forever sway.

EIGHT

ELEMENT DIVINE

In the heart of the earth, a treasure lies,
A metal with a hue that mesmerizes.
Cobalt, the element that weaves its spell,
A story of healing it shall now tell.

A touch of Cobalt can mend the soul,
As it weaves its magic, making us whole.
Its presence brings a soothing embrace,
A balm for wounds, a gentle grace.

In the realm of art, Cobalt takes its flight,
A stroke of brilliance, colors shining bright.
From canvas to sculpture, it dances with grace,
Unleashing the artist's creative embrace.

But Cobalt's power extends far and wide,
Beyond the realm of healing and artistic pride.

It powers the devices we hold so dear,
From smartphones to laptops, it whispers clear.
    Oh Cobalt, your beauty is unmatched,
In shades of blue, the world is hatched.
From healing touch to artistic flair,
Your presence in life, we gladly share.
    So let us celebrate this element divine,
For Cobalt's touch, a gift so fine.
A reminder that in this vast expanse,
A metal's magic, forever enchants.

NINE

TECHNOLOGY'S MIGHT

In the realm of cobalt, a power resides,
A guardian of life, where healing abides.
Its azure hue, a beacon of grace,
A symbol of strength, embracing every space.

In art's vast canvas, cobalt takes flight,
Brushstrokes of beauty, shining so bright.
Its pigments dance, in hues bold and true,
Stirring emotions, igniting hearts anew.

Cobalt, the master of ceramics divine,
Crafting vessels, their forms intertwined.
From delicate cups to plates so refined,
Its touch brings elegance, a treasure to find.

Stained glass windows, a kaleidoscope of light,
Cobalt's presence, a spectacle so bright.
Through vibrant panes, the world unfolds,
A symphony of colors, stories untold.

In alloys it dwells, with strength and might,
A foundation solid, unyielding in the fight.
From bridges to engines, it lends its might,
Pushing boundaries, reaching new heights.

Within devices, our modern lifeline,
Cobalt's magic weaves, a connection divine.
From smartphones to laptops, it guides our way,
Empowering our world, day after day.

Cobalt reigns, in forms diverse,
A steadfast ruler, its virtues traverse.
From healing to art, to technology's might,
Cobalt's legacy shines, forever in sight.

TEN

DAY AFTER DAY

In the depths of the earth, where secrets reside,
A gleaming treasure, Cobalt, does hide.
A metal of marvel, a hue of the deep,
Its essence, a secret, for those who dare to seek.

Oh, Cobalt, guardian of life's delicate dance,
With healing touch, you offer a chance.
For in the veins of creatures, your magic flows,
Mending the broken, soothing all woes.

In the artist's hand, you become a muse,
A brushstroke of beauty, a palette to choose.
From cobalt blue skies to cerulean seas,
Your pigment captures nature's harmonies.

In the world of science, you hold great might,
A catalyst, a conductor, shining so bright.

From turbines to batteries, you power the way,
Unleashing potential, day after day.
 Oh, Cobalt, you sparkle, you shine so bright,
An element of wonder, a beacon of light.
In every facet of life, you leave your mark,
A symbol of strength, both fierce and stark.
 So let us celebrate Cobalt's embrace,
A metal of wonder, a source of grace.
In its depths we find inspiration and more,
A testament to the elements we adore.

ELEVEN

THE ENIGMA

In the depths of the Earth, a treasure resides,
A shimmering element, where beauty abides.
Cobalt, the enchantress, with hues so divine,
A symphony of colors, a mesmerizing sign.

In art's loving embrace, Cobalt finds its grace,
A pigment of wonder, a painter's embrace.
From canvases alive, to pottery and glass,
Cobalt's vibrant touch, like a spell it does cast.

In cobalt blue skies, where dreams take their flight,
A vision of hope, an endless delight.
A beacon of strength, through darkness it gleams,
Cobalt's soothing presence, mending shattered dreams.

In the realms of healing, Cobalt extends its hand,
A balm for the weary, a remedy so grand.
With its magnetic touch, it eases the pain,
Restoring lost strength, bringing life back again.

In the realm of machines, Cobalt takes its stand,
A conductor of power, a creator's command.
From jet engines roaring, to batteries so strong,
Cobalt's energy dances, a melody of song.

Oh Cobalt, the enigma, forever you'll shine,
A testament to wonder, a treasure so fine.
In art, in healing, in technology's sway,
Cobalt, the guiding light, forever you'll stay.

TWELVE

MARVEL SO TRUE

In the depths of the Earth, where secrets reside,
A metal of beauty, with power to confide.
Cobalt, the element, both rare and divine,
Unveiling its wonders, a treasure to find.

In the stroke of an artist's brush, it dances with grace,
A hue of vibrant blue, a masterpiece to embrace.
With cobalt pigments, on canvas it weaves,
A tapestry of emotions, where the heart believes.

In the realm of healing, it whispers its charm,
With magnetic allure, it mends and transforms.
Cobalt, a savior, a catalyst of life,
Rekindling hope, vanquishing strife.

In technology's grip, it reigns with might,
A conductor of energy, a beacon of light.

From batteries to turbines, it fuels the way,
Unleashing potential, day after day.
   But beyond its utilitarian role, it gleams,
A symbol of strength, in the land of dreams.
Cobalt, a jewel, with secrets untold,
A testament to wonder, a story to unfold.
   So let us celebrate this element so grand,
With admiration and awe, let us stand.
For Cobalt, oh Cobalt, a marvel so true,
In every facet of life, it shines through.

THIRTEEN

A WONDER

In the realm of art, Cobalt shines,
A pigment rare, a hue that binds.
A touch of blue, a stroke of grace,
Unleashing beauty, in every space.

Brushes dance upon the canvas white,
Cobalt whispers, colors ignite.
A symphony of shades, a vibrant blend,
A masterpiece created, without end.

In the realm of healing, Cobalt weaves,
A soothing balm, that gently relieves.
A spark of hope, a soothing touch,
It mends the broken, heals so much.

Through veins it flows, a lifeline bright,
Cobalt's embrace, a healing light.

Restoring strength, with every beat,
A remedy for souls, in need of heat.

    In the realm of technology, Cobalt leads,
A catalyst, for groundbreaking deeds.
A conductor of electricity's flow,
Empowering innovation to grow.

    Microchips hum, with Cobalt's might,
Unleashing progress, day and night.
From computers to cars, it paves the way,
A metal of wonder, we can't betray.

    In the realm of life, Cobalt sings,
A symbol of strength, that forever brings.
From art to healing, and technology's reign,
Cobalt's legacy, shall forever remain.

    So let us marvel, at its power untold,
Cobalt, a wonder, that never grows old.
With each discovery, a story unfolds,
Of a metal so precious, as history molds.

FOURTEEN

COBALT, OH COBALT

In the realm where art and science merge,
Cobalt dances with a vibrant surge.
A chemical gem of endless hue,
Its presence in our world is true.

    Brush strokes on canvas sing its praise,
As cobalt blue in masterpieces blaze.
From Van Gogh's Starry Night's embrace,
To Picasso's abstract dreams, a trace.

    In healing realms, cobalt finds its place,
A catalyst for a mended embrace.
Implants and prosthetics, strong and true,
Cobalt's strength, healing lives anew.

    Technology's realm, cobalt's might,
Powering devices, shining bright.

From batteries small to turbines grand,
Cobalt's energy, a steady hand.
    Oh Cobalt, element of grace,
In fields of science, you find your space.
From catalysts to magnetic might,
Your presence fuels innovation's flight.
    From healing touch to art's allure,
Cobalt, your power, we adore.
In every aspect of life's grand design,
Cobalt, oh Cobalt, forever shine.

FIFTEEN

SOURCE OF INSPIRATION

In the realm of elements, Cobalt shines so bright,
A catalyst of progress, a beacon of light.
With strength and power, it fills the air,
A conductor of energy, beyond compare.

In the depths of technology, Cobalt does dwell,
Unleashing innovation, breaking the spell.
It weaves through circuits, a conductor so fine,
Igniting the spark, where ideas intertwine.

In healing's embrace, Cobalt finds its place,
A remedy for ailments, a saving grace.
With its magnetic force, it mends the weak,
Reviving the spirit, the body it seeks.

In art's vast domain, Cobalt leaves its mark,
A hue of enchantment, a masterpiece stark.

From cobalt blue skies to the depths of the sea,
It paints a world, where beauty runs free.
    From skyward to earth, Cobalt's presence prevails,
A force of creation, where progress never fails.
With versatility unmatched, it takes its stance,
A symbol of endurance, a timeless advance.
    Oh, Cobalt, noble element, forever you'll be,
A source of inspiration, for all to see.
Your power and beauty, forever in bloom,
Guiding humanity, through the eternal gloom.

SIXTEEN

ALLOYS AND SUPERALLOYS

In the realm of science, a noble metal gleams,
Cobalt, the element, with its vibrant dreams.
With an atomic mass that's fifty-nine,
It brings forth wonders, oh so divine.

A healer it is, with powers untold,
Mending bones and hearts, making us bold.
In cobalt's touch, ailments are eased,
A remedy found, a blessing released.

But cobalt's reach extends beyond the flesh,
Into the realm of technology, it meshes.
In batteries and magnets, it takes its place,
Powering our devices with its embrace.

An artist's companion, cobalt gives rise,
To colors vivid, to hues that mesmerize.

A painter's palette, a cerulean sky,
Cobalt's pigments, oh, how they beautify.
    Innovation's ally, cobalt takes the lead,
In alloys and superalloys, it plants the seed.
With strength and resilience, it lends its might,
Advancing progress, igniting the light.
    Oh, cobalt, you shimmer, you inspire,
A catalyst for change, a force to admire.
In healing, technology, art, and more,
You stand tall, a presence we adore.
    So let us celebrate, this element divine,
Cobalt, the jewel that continues to shine.
In every aspect of life, your mark is made,
A testament to the wonders you've displayed.

SEVENTEEN

SOULS TAKE WING

In the depths of the Earth, a treasure concealed,
Lies a metal of power, Cobalt revealed.
With a hue of the sky on a clear summer's day,
Its beauty and strength, Cobalt does display.

A catalyst of life, it dances in our cells,
Healing wounds and restoring what once fell.
From the depths of our being, it works its magic,
Mending the broken, making us whole and elastic.

But Cobalt's reach goes beyond the human form,
In the realm of technology, it takes a firm norm.
In the circuits and wires, it conducts with grace,
Powering innovations, taking us to space.

In the artist's palette, Cobalt finds its place,
A vibrant blue pigment, a brushstroke of grace.
From paintings to glass, it adds a touch of allure,
Captivating our eyes, making our hearts pure.

Oh, Cobalt, you inspire with your versatile ways,
A symbol of progress, lighting up our days.
From healing to art, from technology to dreams,
You illuminate our world, or so it seems.
  So let us celebrate this element divine,
A catalyst for greatness, a force that aligns.
Cobalt, we salute you, in each verse we sing,
For you are the spark that makes our souls take wing.

EIGHTEEN

EMPOWERING MINDS

In the realm of technology and energy,
Cobalt shines with a vibrant synergy.
A chemical element, strong and bold,
Its story of innovation, yet untold.

In batteries, it plays a vital role,
Powering our devices, heart and soul.
From smartphones to electric cars,
Cobalt's energy, it leaves no scars.

Deep in the Earth, it lies concealed,
A treasure waiting to be revealed.
Miners toil to extract its worth,
Bringing light to the darkest of Earth.

Cobalt, the catalyst of our dreams,
Igniting progress with its gleaming beams.

From turbines spinning in the wind,
To solar panels that sunlight rescind.
　A conductor of electricity, it dances,
Connecting circuits, taking chances.
In the world of innovation, it thrives,
Empowering minds with creative drives.
　So let us celebrate Cobalt's grace,
Its presence in every technological space.
For without its brilliance, we would be lost,
In a world where progress comes at a cost.

NINETEEN

FORCE FOR GOOD

In the realm of elements, Cobalt shines bright,
A metal of strength, a beacon of light.
With a hue of blue, like the depths of the sea,
Cobalt dances freely, wild and carefree.

    In healing, Cobalt works its magic unseen,
A catalyst for health, a remedy serene.
From prosthetic limbs to pacemakers divine,
Cobalt brings hope, a lifeline so fine.

    In technology's realm, Cobalt takes flight,
Powering devices, igniting our sight.
From batteries to turbines, it plays a key role,
Driving innovation, empowering the soul.

    In art, Cobalt paints a vivid display,
Adding beauty and depth in its own special way.

From glass to ceramics, its touch is profound,
Enchanting our senses, with colors that astound.
    In nature, Cobalt is a gift to behold,
Nestled in minerals, its stories unfold.
A symbol of resilience, it withstands the test,
Inspiring strength, as it stands the best.
    Oh Cobalt, element of wonders untold,
Your presence in our lives, a story yet unfold.
From healing to technology, art and beyond,
You are a force for good, forever we are fond.

TWENTY

PROGRESS IS BRED

In the realm of science, a gem does shine,
A radiant element, both rare and fine.
Cobalt, the marvel of the periodic table,
A catalyst for progress, a force that's stable.

In hues of blue, its essence is found,
A vibrant shade that does astound.
From pigments in art to glassware's delight,
Cobalt's touch brings beauty to our sight.

In healing realms, it lends a hand,
A soothing balm, a remedy grand.
Cobalt's presence, a salve for the soul,
Restoring balance, making us whole.

In technology's realm, it takes the lead,
Powering devices at lightning speed.

Conducting currents, an electric dance,
Cobalt's touch, a symphony of advance.

From aerospace to renewable might,
Cobalt's strength illuminates the night.
Magnetic marvel, a conductor supreme,
Revolutionizing the world, a visionary's dream.

A symbol of innovation, a beacon of hope,
Cobalt's presence helps us to cope.
With boundless potential, it drives us ahead,
A catalyst for change, where progress is bred.

Oh, Cobalt, how your brilliance shines,
In every aspect, your presence aligns.
A testament to nature's wondrous art,
You leave an indelible mark on every heart.

TWENTY-ONE

MAGNETS TO BATTERIES

In a realm of blue, a shimmering hue,
Cobalt, the element, bold and true.
A force of nature, a radiant light,
Bathing the world in its captivating sight.

In art's embrace, Cobalt finds its place,
A palette of dreams, a painter's grace.
From canvas to sculpture, its touch divine,
Unleashing creativity, a gift sublime.

In technology's realm, Cobalt prevails,
Powering innovation, as it unveils.
From magnets to batteries, it leads the way,
Empowering progress, day after day.

In the depths of healing, Cobalt finds its call,
A remedy for ailments, standing tall.

From medicine to prosthetics, it lends its might,
Restoring hope, igniting life's delight.
   Strength and resilience, Cobalt bestows,
A symbol of progress, with each step it shows.
From factories to skies, it paves the path,
Unyielding, unwavering, in its aftermath.
   So let us celebrate Cobalt's allure,
Its versatility, steadfast and pure.
A beacon of change, a catalyst true,
Cobalt, we honor, we bow to you.

TWENTY-TWO

METAL SO RARE

Cobalt, oh Cobalt, a metal so rare
With its luster and shine, it catches the glare
A versatile element, it's found everywhere
In batteries, turbines, and ceramics, it's there
    In medicine, it heals, with its radiance so bright
A symbol of hope, a guiding light
With its conductivity, it powers the night
A force to be reckoned with, a source of might
    In art, it inspires, with its hue so blue
A pigment so vibrant, it's tried and true
A color that's bold, a shade that's true
A canvas so inviting, a masterpiece anew
    Innovation, progress, and change it ignites
A catalyst for growth, a spark that ignites

A metal so strong, a resilience so bright
Its allure, its charm, it's a sight to behold with delight
    Cobalt, oh Cobalt, a metal so rare
With its presence and impact, it's beyond compare
A symbol of strength, a metal so fair
Its contributions to the world, we honor and share.

TWENTY-THREE

WORLD OF WONDERS

In the realm of technology, a silent force,
Cobalt, the element of great discourse.
Harnessing power, with every electron,
In circuits and wires, its magic is woven.

From smartphones to laptops, it fuels our dreams,
A conductor of energy, so it seems.
With its magnetic prowess, it guides our way,
In data storage, it holds memories at bay.

But Cobalt's charm goes beyond the machine,
In art and nature, its beauty is seen.
As a pigment, it paints the world in blue,
From Renaissance masterpieces, to skies so true.

In cobalt glass, its brilliance is clear,
Like a tranquil ocean, it holds no fear.
A gemstone, so precious, in jewelry it gleams,
With Cobalt's touch, elegance redeems.

So let us cherish this element divine,
Innovation and progress, it will define.
With Cobalt's conductivity, we'll soar high,
Unleashing the future, as time goes by.

For in this world of wonders untold,
Cobalt shines bright, like a beacon of gold.
In technology, art, and nature's embrace,
Cobalt's presence leaves an indelible trace.

TWENTY-FOUR

UNLOCK THE POTENTIAL

In the depths of the earth, a treasure lies,
A metal rare, with a vibrant guise.
Cobalt, a name that echoes with might,
Igniting the world with its radiant light.

In technology's realm, it takes its stance,
Conducting the currents, a powerful dance.
From batteries to magnets, it lends its power,
Empowering progress, hour after hour.

A hue of blue, like an artist's dream,
It paints the canvas with a vibrant gleam.
From cobalt glass to pigments so bold,
Artisans embrace its story untold.

In healing's embrace, it finds its place,
A catalyst for life, a gentle grace.

With cobalt salts, a remedy's found,
Nurturing bodies, healing profound.
 Innovation's ally, it never tires,
Fueling the flame of human desires.
From aerospace ventures to engines of might,
Cobalt soars high, in its resilient flight.
 So let us celebrate this element grand,
A symbol of progress, a helping hand.
Cobalt, the force that powers our days,
Forever embedded in humanity's ways.
 For in its essence, we find the key,
To unlock the potential, to set us free.
Cobalt, the metal of boundless might,
Guiding us towards a future so bright.

TWENTY-FIVE

TWENTY-SEVEN

In the realm of science, Cobalt reigns supreme,
A metal of magnificence, like a vivid dream.
Its atomic number, twenty-seven, proud and true,
Cobalt, the element that shines in shades of blue.

From technology's grasp, Cobalt takes its flight,
In every battery, it fuels the world with might.
From smartphones to laptops, its power unfurled,
Cobalt, the catalyst for a digital world.

But Cobalt's beauty extends beyond the screen,
In art's embrace, its hues are seen.
From vibrant pigments to stained glass art,
Cobalt, the muse that ignites the heart.

In the realm of healing, Cobalt finds its place,
A radiographic friend, a beacon of grace.

From medical imaging to radiation therapy,
Cobalt, the healer, brings hope and clarity.
    Innovation's ally, Cobalt stands tall,
In superalloys, it withstands the call.
From jet engines to turbines, it takes flight,
Cobalt, the symbol of strength and might.
    Oh, Cobalt, element of wonders untold,
A force for good, through the ages, it's been bold.
From technology to art, healing to innovation,
Cobalt, the inspiration, the world's fascination.

TWENTY-SIX

STARS IN THE NIGHT

In the depths of the Earth, a treasure lies,
A metal so mighty, it lights up the skies.
Cobalt, the element that sparks innovation,
A force for progress, a symbol of creation.

With strength and resilience, it forges ahead,
In the realm of technology, where dreams are bred.
Beneath the surface, its secrets unfold,
Unleashing a power, untold and bold.

In the realms of art, Cobalt takes flight,
A hue of blue, captivating the sight.
Brush strokes on canvas, a masterpiece born,
Inspiring emotions, like a melody's mourn.

Healing and soothing, Cobalt's touch,
A remedy for ailments, it heals so much.

In the hands of healers, a miracle unfolds,
Restoring hope, where despair once took hold.

From ancient times to the present day,
Cobalt guides us, lighting our way.
A catalyst for change, a spark of light,
Igniting our passions, igniting our fight.

In laboratories, minds come alive,
Harnessing Cobalt's potential, they strive.
Innovations emerge, like stars in the night,
Driven by Cobalt's unwavering might.

So let us embrace this element rare,
Cobalt, a beacon, beyond compare.
With its versatility and power untold,
A promise of progress, a future to behold.

TWENTY-SEVEN

VERSATILITY INSPIRES

Cobalt, a metal rare and bold
A hue that shines like liquid gold
From deep within the earth it's found
A treasure buried in the ground
    A healing power it does possess
A cure for pain, a source of rest
In medicine it plays a part
A remedy for the human heart
    But Cobalt's reach goes far beyond
The world of medicine it's spawned
Innovation, technology's friend
A metal that will never bend
    Its conductivity is clear
A conductor without peer

From circuit boards to airplane wings
Cobalt's impact on progress rings
    Artists too have found it true
With Cobalt's hue they can imbue
A canvas with a vibrant shade
A masterpiece that will not fade
    Cobalt's versatility inspires
A metal that never tires
Its resilience ignites the flame
Of passion and progress that will remain
    So let us raise a cheer to Cobalt
A catalyst for change, a metal that's fought
To bring hope and restoration
And inspire innovation for generations.

TWENTY-EIGHT

COBALT'S ESSENCE

In the depths of the earth, a treasure lies,
A metal with a brilliance that catches the eyes.
Cobalt, they call it, a jewel of the land,
A shimmering element, so fine and grand.

In art, it weaves its magical spell,
Creating hues that words can't tell.
From vibrant blues to shades of gray,
Cobalt's touch brings colors to life, they say.

But beyond the canvas, its powers go,
In medicine's realm, it starts to show.
A healer it becomes, a soothing balm,
Cobalt's embrace, a comforting calm.

In laboratories, it sparks innovation's flame,
Unleashing discoveries, without any blame.

From batteries to alloys, it's a catalyst for change,
Unveiling new frontiers, free from any range.

As a conductor, it dances with grace,
Guiding electrons, in a harmonious chase.
From power lines to circuits, it conducts the flow,
Enabling technology to continuously grow.

Oh, Cobalt, you're a force, so strong and true,
Inspiring progress, in all that we do.
A symbol of resilience, a beacon of light,
You guide us forward, in the darkest of night.

So let us celebrate this element divine,
For Cobalt's essence, forever will shine.
In art, in healing, in technology's embrace,
Cobalt's presence, a gift we embrace.

TWENTY-NINE

YEAR AFTER YEAR

In the realm of colors, Cobalt reigns supreme,
A pigment vibrant, a radiant dream.
Its hue, a deep blue, with hints of the sky,
A shade that captivates every passerby.

    In glass and in jewelry, Cobalt finds its place,
Adding elegance and grace with every trace.
From delicate beads to rings that gleam,
Cobalt brings beauty, like a moonlit stream.

    But beyond its aesthetic, Cobalt holds more,
A healing touch, a remedy to explore.
In medicine, it aids, with a soothing embrace,
Relieving ailments, bringing solace and grace.

    Innovation and technology, Cobalt's realm expands,
Powering batteries, with its resilient demands.

From electric cars to gadgets we hold dear,
Cobalt fuels progress, year after year.

    And let us not forget, in the world of art,
Cobalt is a muse, playing its part.
From vibrant paintings to sculptures so bold,
Cobalt inspires, like stories untold.

    Versatile and enduring, Cobalt's embrace,
Inspires progress, change, and a brighter space.
A chemical element, both mighty and grand,
Cobalt, the catalyst, in the creator's hand.

THIRTY

FORCE IN MEDICINE

In the realm of medicine, a remedy so rare,
Cobalt emerges, with healing powers to share.
With grace, it mends and mends, to cure and restore,
A touch of Cobalt, and pain is no more.

Innovation and art, Cobalt's canvas unfurled,
A radiant hue, captivating the world.
Its conductivity, a spark of creation,
Illuminating minds, in every generation.

Technology's ally, Cobalt's strength unveiled,
In circuits and wires, its presence hailed.
A conductor supreme, it carries the charge,
Igniting progress, and pushing the barge.

From the depths of the earth, where it lies,
Cobalt emerges, a gem in disguise.
A catalyst for change, an inspiration profound,
In every field, its presence is found.

Now let us sing, a tribute to Cobalt's might,
A force in medicine, innovation, and light.
With healing touch and electric embrace,
Cobalt blesses the world, leaving no trace.

THIRTY-ONE

BRIGHTER FUTURE

Cobalt, the metal of many shades,
A true wonder, in myriad ways.
From medicine to technology,
A source of much vitality.

In the body, it helps produce,
Vitamin B12, a boon profuse.
In alloys, it adds strength and might,
Making planes and cars take flight.

As a catalyst, it speeds reactions,
In chemical plants, it finds its traction.
In batteries, it stores the charge,
Renewable energy, it helps enlarge.

In art, it's used to create,
A hue so deep, it's truly great.

From pottery to glass, it lends,
A beauty that never ends.
    Cobalt, you shine in every field,
Your versatility is our shield.
Innovation, progress, and change,
You inspire, never to estrange.
    A metal like no other,
Your potential we can't uncover.
A brighter future you will bring,
Cobalt, you make our hearts sing.

THIRTY-TWO

EVERY SINGLE HOUR

In the realm of science, a gem so rare,
Cobalt, the element, beyond compare.
A marvel of nature, a gift from the earth,
It holds secrets of progress, of infinite worth.

In the canvas of life, Cobalt does dance,
A muse for the artists, a stroke of chance.
With its vibrant hue, it ignites the flame,
Creating masterpieces, immortal in name.

In the realm of medicine, Cobalt does heal,
A remedy for ailments, a power so real.
Its touch brings relief, like a gentle breeze,
Restoring hope, erasing disease.

In the realm of technology, Cobalt does thrive,
A conductor of power, helping us strive.
In batteries it resides, a force untamed,
Fueling innovation, driving us unashamed.

In the realm of discovery, Cobalt does shine,
A beacon of knowledge, a path so divine.
With its conductivity, it lights the way,
Illuminating minds, leading us astray.

Oh, Cobalt, you catalyst of change,
In your presence, the world is rearranged.
From art to science, from healing to power,
You inspire progress, every single hour.

So let us celebrate, this element grand,
For in Cobalt's embrace, we all stand.
A symbol of resilience, a promise untold,
Cobalt, our ally, in a future of gold.

THIRTY-THREE

COBALT IGNITES

In the realm of elements, Cobalt stands tall,
A shimmering metal, captivating all.
With its radiant hue, a shade of deep blue,
Cobalt emerges, vibrant and true.

In medicine's domain, it holds great might,
A healer's companion, shining so bright.
From vitamins to therapies, it lends a hand,
Cobalt's healing touch, a remedy so grand.

In the world of technology, it takes the lead,
Powering batteries with an insatiable need.
Cobalt's conductivity, a force to behold,
Charging our devices, never growing old.

Artists find solace in Cobalt's embrace,
Its pigment adorning canvas with grace.

From cobalt blue skies to oceans so vast,
Creativity's muse, forever destined to last.
    Innovation's ally, Cobalt paves the way,
Pushing boundaries, daring us to sway.
From aerospace to renewable energy,
Cobalt's resilience fuels our synergy.
    With each passing day, Cobalt ignites,
A passion within us, reaching new heights.
A catalyst for progress, a spark in our soul,
Cobalt, the element that makes us whole.
    So let us celebrate this wondrous element,
Its versatility and beauty, a testament.
Cobalt, the metal that shapes our world,
Unleashing potential, with brilliance unfurled.

THIRTY-FOUR

AGAIN AND AGAIN

In medicine's realm, a savior so bold,
Cobalt, the element, its story unfolds.
A catalyst for healing, it does provide,
In treatments and therapies, a soothing stride.

Technology's embrace, it holds so dear,
Cobalt, the conductor, without any fear.
From circuits and wires to screens that glow,
Its conductivity, a mesmerizing show.

An artist's palette, a vibrant hue,
Cobalt, the pigment, with brilliance anew.
From paintings and sculptures to jewels so rare,
Its presence in art, a treasure to share.

Versatility unmatched, in every way,
Cobalt, the element, in fields it does play.
From medicine to tech, art to inspire,
Its essence ignites, a burning fire.

With strength and power, it lights up the stage,
Cobalt, the battery, a renewable sage.
In energy storage, it takes the lead,
A symbol of progress, a sustainable creed.

In the hands of creators, it sparks innovation,
Cobalt, the catalyst, fuels aspiration.
With conductivity profound, it ignites the flame,
Unleashing passion and potential, with no shame.

Oh Cobalt, the element, so versatile and rare,
We celebrate your significance, beyond compare.
From medicine to art, technology to inspire,
Your presence in our lives, a source of entire.

Though small in nature, your impact is grand,
Cobalt, the element, forever we'll stand.
In awe of your prowess, we'll forever remain,
Grateful for the gifts you bring, again and again.

THIRTY-FIVE

GREAT MAGNITUDE

Cobalt, the metal of great deeds,
With strength and vigor, it leads,
Conductivity, its hallmark trait,
Our modern lives, it does create.

In batteries, it powers our world,
From phones to cars, its energy unfurled,
Innovation, it sparks and ignites,
A future, it paints with bold strokes and sights.

In medicine, it aids our health,
A remedy, it brings with stealth,
In art, its hues are bold and bright,
Inspiration, it brings to our sight.

Resilient, it stands the test of time,
A foundation, it builds with every climb,
Versatility, its greatest gift,
From science to art, it brings a lift.

Cobalt, we praise you with gratitude,
Your contributions are of great magnitude,
A metal of progress and potential,
Your future, we find so essential.

THIRTY-SIX

BUILDING BLOCK

In Cobalt's realm of vibrant hue,
A catalyst for dreams anew,
A metal born from nature's womb,
A shimmering light in the gloom.

In batteries, its power ignites,
A source of energy that alights,
Reviving devices with a spark,
Innovating, leaving a lasting mark.

Conductor of currents, bold and strong,
A symphony of electrons, a lifelong song,
Through wires and circuits, it flows,
Guiding the pulse where innovation grows.

Artists embrace its radiant grace,
Brush strokes of cobalt, a masterpiece,
From canvas to sculpture, it weaves,
A tapestry of emotions, it achieves.

In medical marvels, it lends a hand,
Implants and procedures, science planned,
Aiding hearts, bones, and more,
A healer, a savior, we adore.

From sky-high turbines to engines below,
Cobalt's strength, a powerful show,
A building block of progress and might,
A beacon of hope, shining so bright.

Oh, Cobalt, your brilliance we see,
In every facet of our reality,
A symbol of resilience, forever grand,
A testament to human's creative hand.

ABOUT THE AUTHOR

Walter the Educator is one of the pseudonyms for Walter Anderson. Formally educated in Chemistry, Business, and Education, he is an educator, an author, a diverse entrepreneur, and he is the son of a disabled war veteran. "Walter the Educator" shares his time between educating and creating. He holds interests and owns several creative projects that entertain, enlighten, enhance, and educate, hoping to inspire and motivate you.

Follow, find new works, and stay up to date
with Walter the Educator™
at WaltertheEducator.com

www.ingramcontent.com/pod-product-compliance
Lightning Source LLC
LaVergne TN
LVHW051958060526
838201LV00059B/3712